Fuse Protection against the Regenerative Circuit Internal Commutation Fault

By
Dr. Hidaia Mahmood Alassouli
Hidaia_alassouli@hotmail.com

1. Introduction:

Current-limiting fuses are widely used to protect the thyristors in dc drive systems. In the event of the commutation failure when regenerating (inverting), the fuses need to interrupt in loop supplied by the ac and dc voltages acting in series which is the most difficult case for protection by fuse.

In this paper, a detailed study of the complete interruption process has been investigated by modeling of arcing process of the fuse. The effect varying the motor time constant, supply impedance, number of fuses used to clear the fault and dc machine rating to the total response for fuses protecting against the regenerative circuit internal commutation fault has been studied. The model of 200A fuse is employed in this study and fuses in series with the semiconductor devices (F1) and fuses in ac line (F2) are both considered.

2. The main model used in the study:

The basic function of the phase-controlled converter is to convert an alternating input voltage to controllable direct output voltage. The 2-qudrant converter consists basically of a conventional rectifier circuit arrangement which contains thyristors. Due to the unidirectional current carrying property of the thyristors, the current at dc terminal can flow only in one direction. However, it is possible by suitable control of the phase position of the firing pulses applied to the thyristors with respect to ac input voltage, for the mean voltage at dc terminals to be continuously controlled from maximum positive to maximum negative. Thus power can flow either from ac to dc side of the converter, and vice versa, and the converter is capable to work as an inverter or rectifier.

The average output voltage is given by $V_d = 1.35 E_{LL} \cos a$. When 0<α< 90, V_d positive (rectifier mode), while when 90<α< 180, V_d negative (inverter mode). If the 2-quadrant converter is connected to the armature of a dc machine, the polarity of the induced emf of the machine reverses whenever regenerative breaking is required. For example, when the load is being raised, the converter supplies power to machine and induced emf is positive. On the other hand, when the load lowered, the direction of rotation and induced emf are reversed. Therefore, the polarity of the machine emf is now negative. Consequently, it is in the correct direction for the converter to operate in its inverter region, returning power from machine to ac system.

Internal commutation fault will occur if a thytristor that has being conducting fails to turn off. Alternatively, this fault might occur if any thyristor fails to turn on at specified point when the thyristor assumed to conduct.

If two thyristors conduct simultaneously, then the line voltage will be applied to the load. Assuming the first instance that the commutation fault is due to loss of trigger pulses to all thyristors, then, the circuit in this condition

$$e_{ac} + E_{dc} = (R_{ac} + R_{dc})i + (L_{ac} + L_{dc})\frac{di}{dt}$$

The fuse is usually several parallel elements and several series notches per element and can be modeled as cylindrical plasma. The fuse arc voltage during arcing in this study was represented by the following simplified 5th order dynamic model. When the fuse arc voltage is considered during the arcing period:

$$\frac{di}{dt} = \frac{\sqrt{2}E_{LL}\sin(\omega t + \theta) + E_{dc} - Ri - V_f}{L}$$

The arc voltage is given by

$$V_f = \begin{cases} 0 & (prearcing) \\ xE & (arcing) \end{cases}$$

$$\frac{d(i^2 t)}{dt} = i^2$$

$$\frac{d(arc-energy)}{dt} = V_f i$$

$$\frac{dx}{dt} = C_s i$$

$$\frac{dA}{dt} = \gamma\alpha \frac{E(i_e, A) i_e}{H_f}$$

$$E = \frac{K_1 . i_e^{0.4}}{A^{0.85}} + \frac{K_2 . i_e^{-0.4294}}{A^{0.5}} \qquad K_1 = 4.58 \qquad K_2 = 46.8$$

$$\alpha = \alpha_0 + (\alpha_m - \alpha_0)(1 - e^{-t/\tau})$$

$$\alpha_o = 0.0655 \qquad \alpha_m = 0.63 \qquad \tau = 0.01$$

Where, N_p: Number of parallel elements =2, $i_e = i / N_p$: Current per element, A: Effective cross-sectional area of lumen (arc channel cross section), x: Total length of all arcs in series, H_f: Fusion enthalpy of sand ≈ 3.614, γ: Filler flow multiplying factor ≈ 0.7

As initial values for solving the differential equations, assuming the notch zones will produce a jump in resistance from zero to R_o. If the current at the start of arcing is i_o, the voltage produced is $v_o = i_o R_o \qquad R_o$ » 0.05

If the initial arcs channel area A is assumed same as the strip cross section S, then for each sub element the arc gradient will be

$$E_o = E(\frac{i_o}{N_p}, A_o)$$

And the arc length that will satisfy this will be $x_o = \frac{\lfloor i_o R_o \rfloor}{E_o}$

The transition from pre-arcing to arcing state occurs when $i^2 t$ integral reaches the melting value. However in practice, the pre-arcing $i^2 t$ increases according to

$$|i^2 t| = |i^2 t|_{\min}(1 + 345t)$$

For the dc side of the converter, the volt-ampere is

$$S_{dc} = V_d I_{dc}$$

For the ac side, the volt-amperes and line current are

$$S_{ac} = \sqrt{3} E_{LL} I_L, \; I_L = \sqrt{\frac{2}{3}} I_{dc}, \; V_d = 1.35 E_{LL}, \; S_{ac} = 1.05 V_d I_{dc}, \; I_{dc} = \frac{Pe}{E_{dc}}$$

When I_{dc} is calculated the minimum permissible fuse rating (MPFR) is evaluated. There are always two fuses in series, ether in series with the semiconductor devices or in the ac line. The paper will investigate both locations. The fuse sizes should be at least 125% of the rms load current. Therefore, the minimum permissible fuse ratings (MPFR) for both locations are

$I_{F1} = 1.25\frac{1}{\sqrt{3}}I_{dc}$, $I_{F2} = 1.25\frac{2}{\sqrt{3}}I_{dc}$

Then the fuse ampere rating (FAR) and the melting $i^{t}t$ suitable for the application are found. The model of 200A fuse will be employed in this study. E_{LL}=460 V, frequency=60 Hz, R_{dc}=0.05 p.u, $E_{dc} = 500V$. Fuses in series with the semiconductor devices (F1) and fuses in ac line (F2) will be considered. If fuse is used in location F1, that gives certain dc loading condition $I_{dc} = \sqrt{3}.1.25I_{F1}$. If fuse is located at F2, this will give a different dc loading condition, $I_{dc} = \sqrt{3}.1.25I_{F2}/2$. The loading conditions for the two systems:

Location	DC machine (HP)	DC full Load current (A)	Converter rating (KW)	Fuse ampere rating (A)
F1	170	127	270	200
F2	120	90	190	200

3. Overall system transient response:

Fig. 1 shows the internal commutation fault current and the total supply voltage $\sqrt{2}E_{LL}\sin(\omega t+\theta)+E_{dc}$ when no fuse was used to clear the circuit. This is the prospective fault current that the fuses have to interrupt. Fig. 2 shows the transient response for total supply voltage, prospective current, fuse voltage, arc energy, i^2t and arc length x. The figures obtained for 0.05 p.u supply impedance and 40msec dc machine time constant.

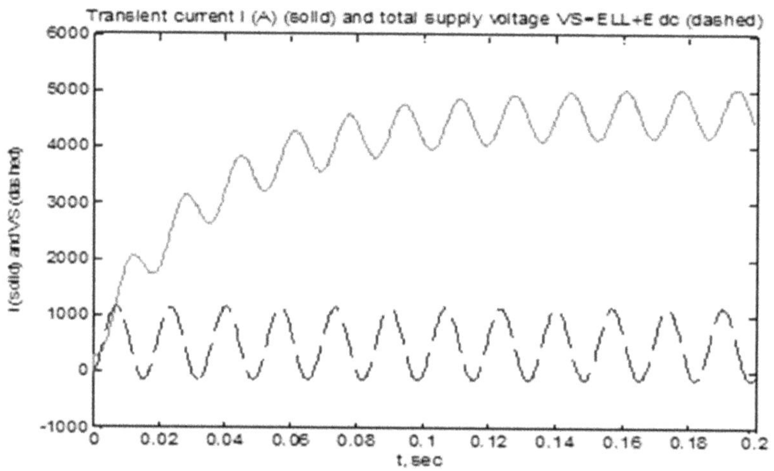

Fig. 1. The internal commutation fault current and the total supply voltage $\sqrt{2}E_{LL}\sin(\omega t+\theta)+E_{dc}$ when no fuse was used to clear the circuit.

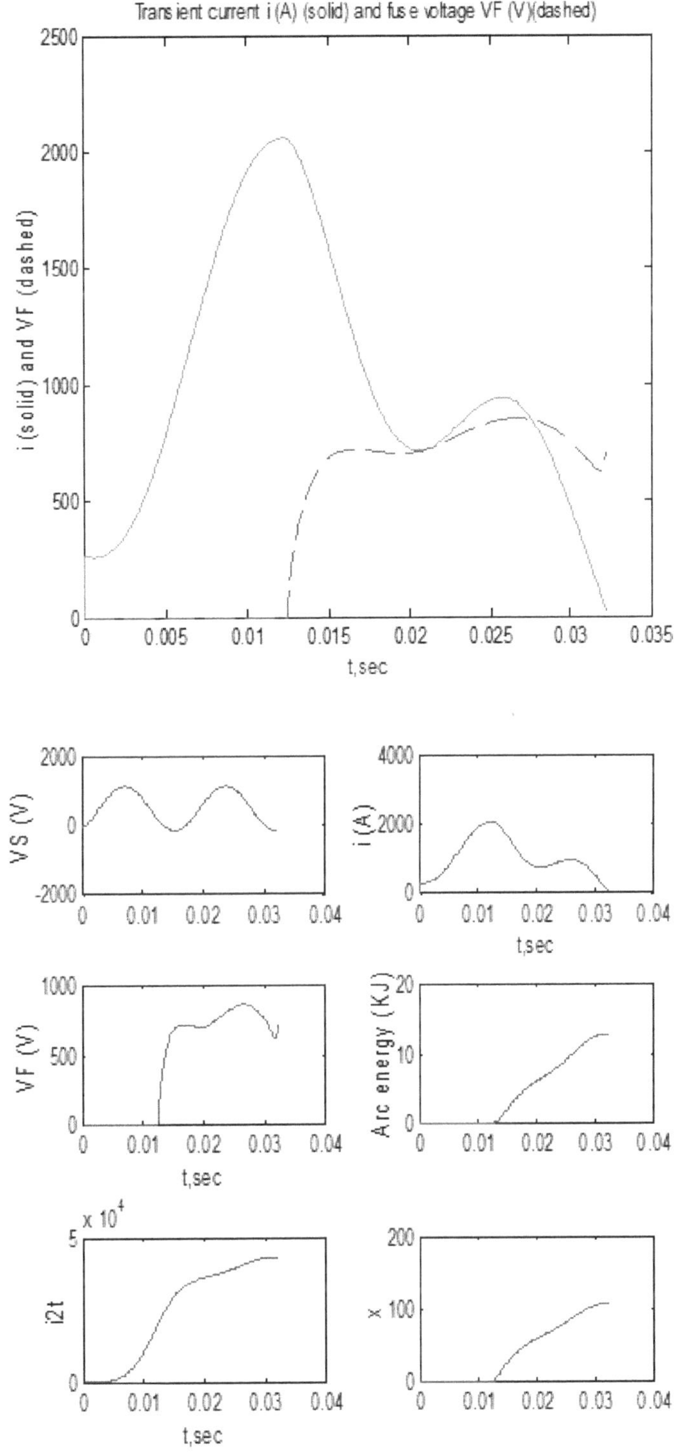

Fig. 2. The transient current and fuse voltage if only one fuse in series with the semiconductor device at F1 was used to clear the circuit.

4. The effect of the motor time constant

The total response to an internal commutation fault depends critically upon the dc load characteristics. The effect of varying the dc machine time constant has been investigated within the range of 1 to 100msec. Fig. 3 shows the variation of the arc energy (KJ) and fuse voltage versus the dc machine time constant when the 200 A fuse used as F1 and F2. Fig. 4 shows the variation of the current at start of arcing (A), peak arcing current (A), pre-arcing time(ms) and arcing duration (ms) versus the dc machine time constant used when the 200 A fuse was used in both locations. Fig. 5 shows the variation of the arc energy versus arc-angle and arc-angle versus the dc machine time constant when the 200 A fuse used as F1 and F2. It is clear that increasing the time constant will increase the arc-angle, pre-arcing time and arcing duration. The minimum range for peak fuse voltage, arc energy, starting and peak prospective currents will occur at the mid range of motor time constant starting from 20msec to 50msec.

Table. 1 lists the minimum and maximum values for arc energy, peak fuse voltage, the current at start of arcing , peak prospective current, pre arcing time, arcing duration, arc-angle and the time constant at which the minimum and maximum occurred.

	Minimum when fuse used as F1	Minimum when fuse used as F2	Maximum when fuse used as F1	Maximum when fuse used as F2
Arc energy (KJ)	9.483 at tc=25 at arc-angle=143.7	7.8012 at tc=15 at arc-angle=146.0	27.857 At tc=100 at arc-angle 739.9205 (19)	32.99 at tc=100 at arc-angle= 1292.57(212.6)
Peak fuse voltage during arcing (V)	839.066 at tc=35	786.11 At tc=20	1135.19 At tc=5	1058.67 at tc=95
Current at start of arcing (A)	1564.93 at tc=50	1431.82 At tc=30	3362.02 at tc=1	3188.17 at tc=1
Peak current (A)	1763.0584 at tc=50	1761.5082 At tc=85	3572.9581 at tc=1	3359.3621 at tc=1
Pre-arcing time (ms)	4.39 at tc=1	5.19 At tc=1	37.03 at tc=100	62.6193 at tc=100
Duration of arcing (ms)	7.0474 At tc=5	6.7122 At tc=2	35.5435 At tc=80	43.2209 at tc=85
Arcing angle	35.0123 At tc=1	52.2041 At tc=1	739.9(19) at tc=100	1292.5779 (212.5779) At tc=100

Table. 1

When the fuse in AC line (F2) it will reach higher level of arc-angle compared when the fuse with semiconductor devices. Generally, the most desirable range of arcing angle for fuse to produce melting is between 90-270, or 450-630 (i.e. falling part of ac voltage). If the arc-angleγ falls within the range 0 to 90 degree, the fuse will melt on the rising part of voltage waveform. If fuse melts within 60 and 90, arcing will occur during a period of time when ac supply voltage near the peak value. This range ofγ will be called critical melting zone. There is another critical melting zone between 420 and 450. Melting mostly will occur in the first critical zone when motor time constant less than 10 ms, and to obtain melting in the second critical zone the time constant shall be increased beyond 50ms, especially from 52-58ms. This is found for both fuse location.

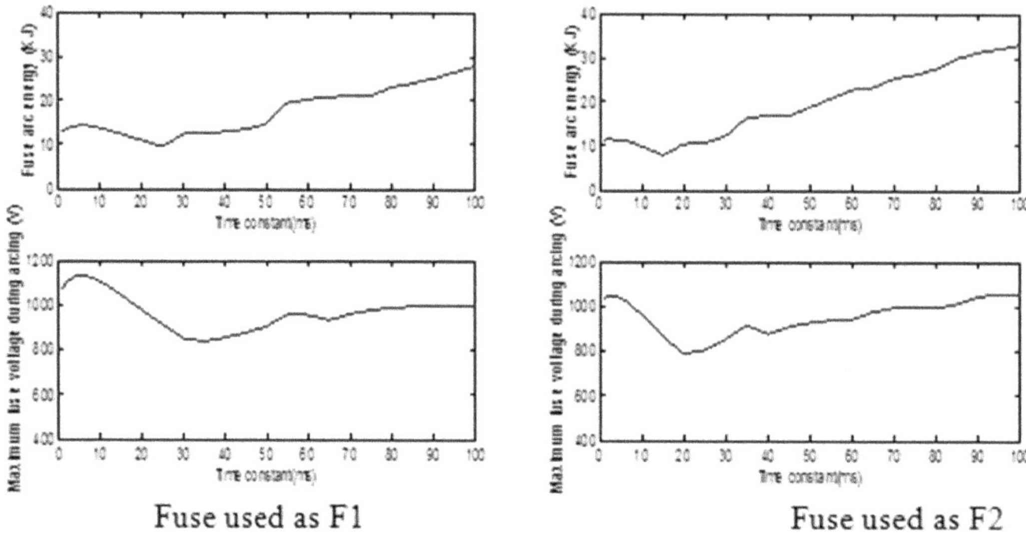

Fig. 3. Variation of the arc energy (KJ) and peak fuse voltage (V) versus the dc machine time constant (msec) when the 200 A fuse used as F1 and F2

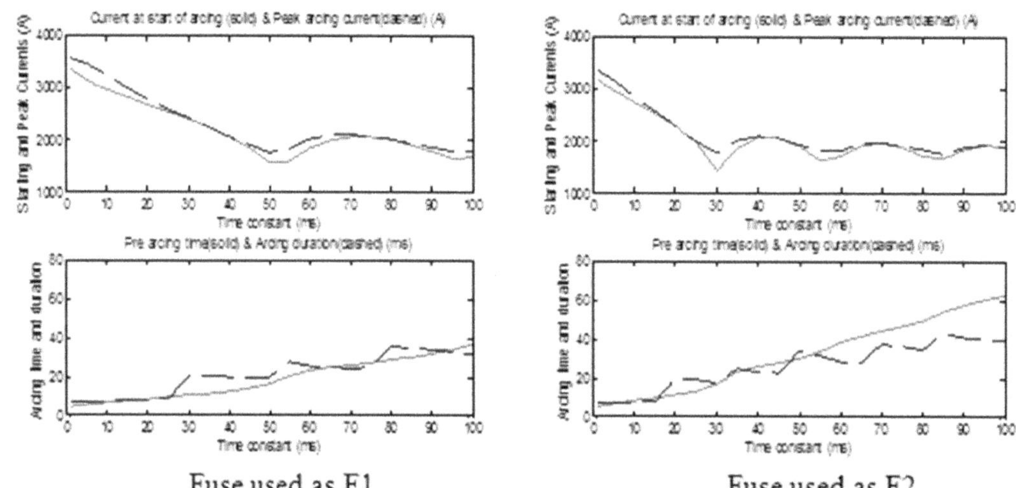

Fig. 4. Variation of the current at start of arcing (A), peak arcing current (A), pre arcing time(ms) and arcing duration (ms) versus the dc machine time constant (msec) when the 200 A fuse used as F1 and F2

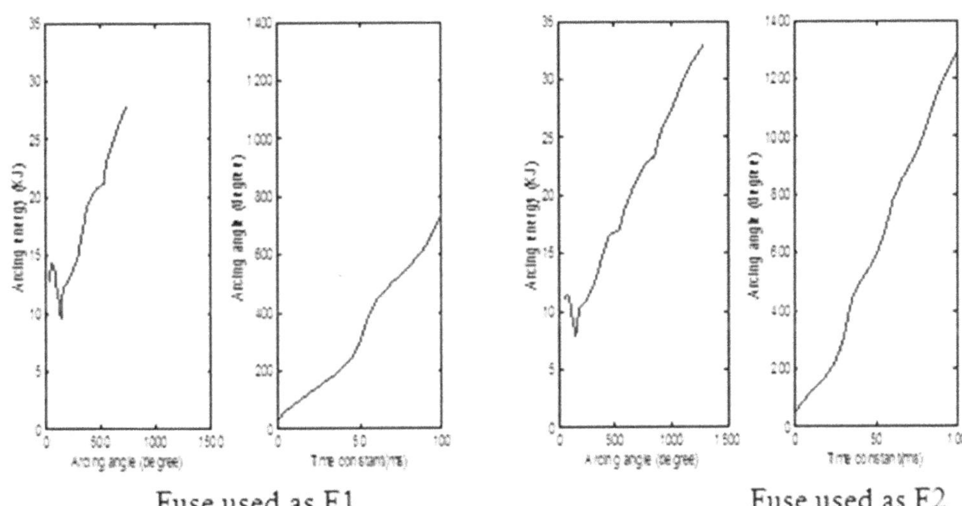

Fig. 5. Variation of the arc energy (KJ) versus arc-angle (degree) and arc-angle (degree) versus the dc machine time constant (msec) when the 200 A fuse used as F1 and F2

5. The effect of number of fuses

It is recommended that that two fuses in series clear the fault. Treatment of two fuses in series sharing the breaking duty in a ratio of f:(1-f) , where f=0.5 and f=0.6 was carried out in this study, where f is the fraction of the total energy or voltage taken by one of the two fuses. The governing equation in this case,

$$\frac{di}{dt} = \frac{V_s - Ri - \frac{1}{f}V_f}{L}$$

Voltage across fuse 1= V_f and across the fuse2 = $\frac{(1-f)}{f}V_f$

Share of total calculated energy=f for fuse 1, and (1-f) for fuse 2. The comparison here was made for a motor with time constant 40 ms and supply impedance 0.05 p.u when having

- One 200 A fuse located at F1 (f=1)
- One 200A fuse located at F2 (f=1)
- Two 200 A fuses located at F1 and sharing the breaking duty equally (f=0.5)
- Two 200 A fuses located at F1 sharing the breaking duty in a ratio 0.6:0.4 (f=0.60)

Table. 2 lists the arc energy, peak fuse voltage, current at start of arcing , peak arcing current, pre-arcing time, arcing duration, arc-angle and total arc energy for each case when the motor with time constant=40 ms and supply impedance=0.05.

	One 200A fuse used as F1, f=1	One 200 fuse used as F2, f=1	Two 200A fuses used as F1, f=0.5	Two 200A fuses used as F1, f=.6
Arc energy (KJ)	12.8722	16.7950	2.0609	3.3757
Peak fuse voltage during arcing (V)	857.6595	876.6067	629.3049	654.6958
Current at start of arcing (A)	2060.5463	2065.6563	2060.5463	2060.5463
Peak current (A)	2062.2832	2100.6173	2062.2832	2062.2832
Pre-arcing time (ms)	12.5337	25.8190	12.5337	12.5337
Duration of arcing (ms)	19.7847	23.1904	15.9757	16.8648
Arcing angle	210.7284	497.6897 = 137	210.7284	210.7284
Arc energy (KJ)	12.8722	16.7950	4.1219	5.6262

Table. 2

It can be noted; when there are one fuse alone clears the circuit, the arc energy and peak fuse voltage during arcing produced for internal commutation fault is high. When there are two fuses either sharing the breaking duty equally (f=0.5) or the voltage is split in the ratio f=0.6, the arc energy and peak fuse voltage is reduced significantly. The pre-arcing time, current at start of arcing, peak prospective current and arc angle were not affected by the number of fuses.

6. The effect of supply impedance:

The percentage value of supply impedance plays an important role for total response of the commutation fault. Here, the supply impedance was varied from X=0.01 p.u to X=0.4 p.u.

Table. 3 lists the maximum and minimum values for arc energy, fuse voltage, the current at start of arcing , peak prospective current, pre-arcing time, arcing duration, arc-angle and the system impedance at which the maximum and minimum occurred when 200A fuse located at F1 and the motor time constant is 40 msec. Their values when the system impedance = 0.05 and 0.15 are listed also. Fig. 6 and Fig. 7 show the variations as the system impedance varied from Z=0.01 p.u to Z=0.4 p.u.

	Z=0.05 p.u	X=.15 p.u.	Minimum when fuse used as F1	Maximum when fuse used as F1
Arc energy (KJ)	12.8722	12.5666	12.5666 at Zpu=0.15 at arc-angle 272	15.6684 at Zpu=.25 at arc-angle 437.47
Peak fuse voltage during arcing (V)	857.6595	857.1913	831.6888 at Zpu=0.35	900.6567 at Zpu=0.25
Current at start of arcing (A)	2060.5463	1623.6856	1425.9338 at Zpu=0.2	2180.4415 at Zpu=0.01
Peak current (A)	2062.2832	1805.4284	1701.1453 at Zpu=0.2	2180.5270 at Zpu=.01
Pre-arcing time (ms)	12.5337	15.3723	11.8 at Zpu=0.01	28.1323 at Zpu=0.4
Duration of arcing (ms)	19.7847	18.610	18.6103 at Zpu=.15	28.0284 at Zpu=0.2
Arcing angle	210.7284	272.0421	195.2552 At Zpu=0.01	547.6584 at Zpu=0.4

Table. 3

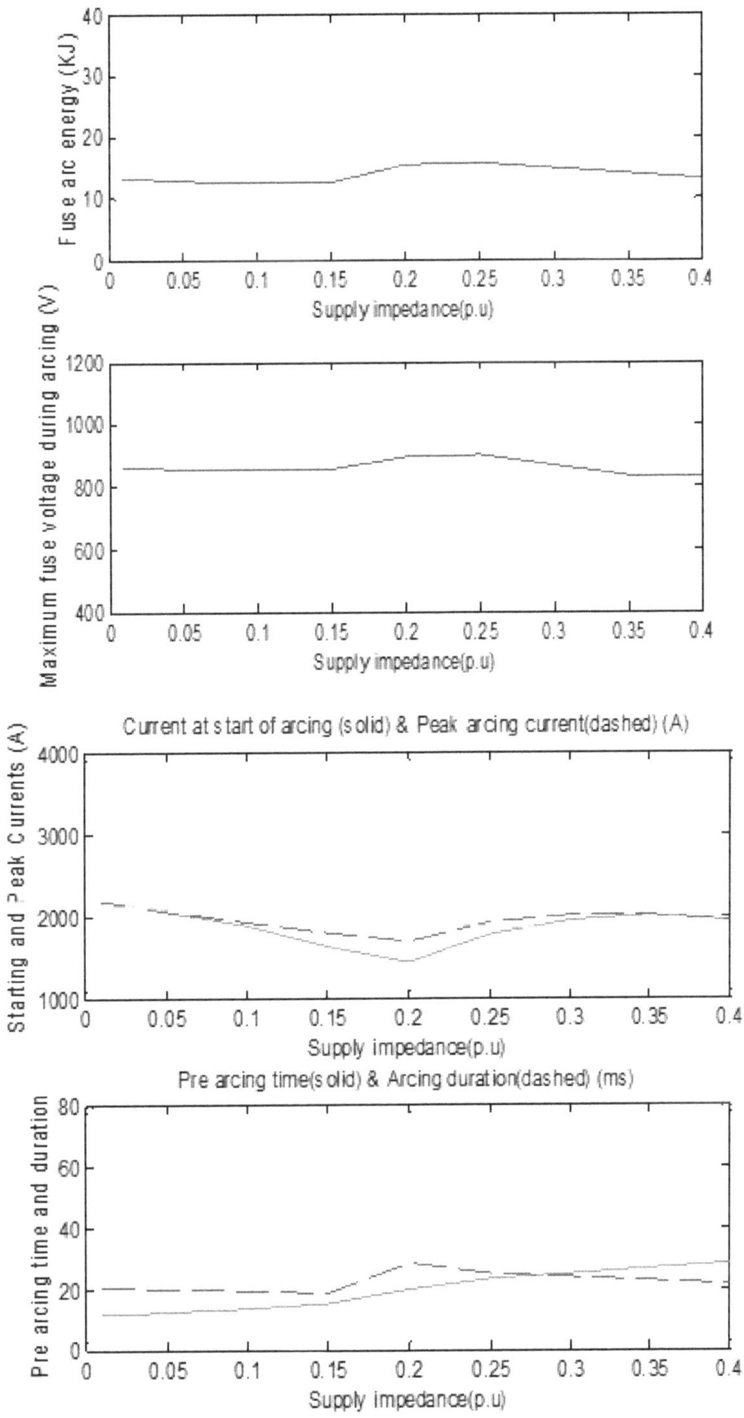

Fig. 6. Variation of the arc energy (KJ) and fuse voltage (V) , current at start of arcing (A), peak arcing current (A), pre arcing time(ms) and arcing duration (ms) versus the supply impedance when the 200 A fuse used as F1

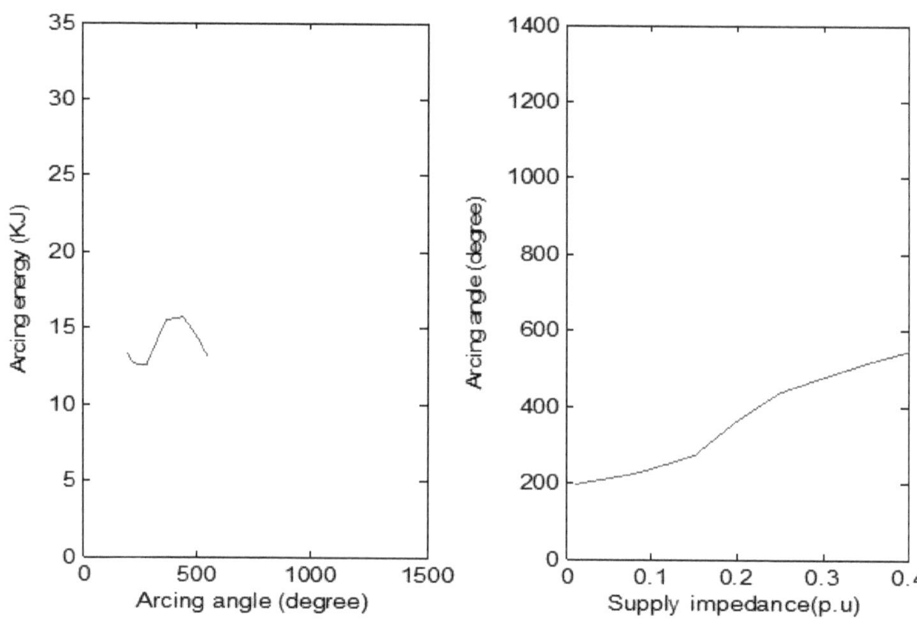

Fig. 7. Variation of the arc energy versus arc-angle and arc-angle versus the supply impedance when the 200 A fuse used as F1

It can be noted that increasing the system impedance will not affect the arc energy or peak fuse voltage during arcing. The arc angle and pre-arcing time increase with the supply impedance in very small range. The current at start of arcing and peak prospective current will be reduced as system impedance increased, but will increase again as Z>0.2 pu.

7. The effect of dc machine rating

The above analysis was done for a range of dc machines with different power rating from 25 HP to 625 HP with the fuse located in series with the semiconductor device (F1). The converter rating in KW will be machine HP*746. The machine time constant is 40msec and supply impedance=0.05 p.u were considered. Table. 4, Fig. 8 and Fig. 9 summarize some of the results. It can be seen that increasing the machine and fuse ratings will increase the level of fuse voltage, starting and peak prospective currents and arc energy to higher level. The arc-angle and pre arcing time generally will not be affected. The arcing duration will decrease sharply as the machine rating increases in the lowest range of machine rating. An adequate voltage rating for fuses protecting the regenerative dc drives can be the ac line-line supply voltage.

Motor HP	Con. Rat (KW)	I_{dc}	MP FR I_{F1}	FAR	Melting i^2t	Arc-angle (degree)	Pre-arc time (msec)	Current at start of arcing (A)	Peak current (A)	Peak voltage during arcing (V)	Arc energy (KJ)
25	18.6	39	28	35	120	342.0	18.6	249.9	402.27	633.22	6.663
50	37.3	79	57	60	360	219.2	12.92	600.1	604.394	594.54	8.157
100	74.6	157	113	125	1600	244.4	14.09	1154.2	1199.17	722.06	11.22
150	111.	235	170	175	3100	214.5	12.70	1790.6	1794.95	826.52	11.81
170	127	270	194	200	4000	210.7	12.53	2060.5	2062.28	857.65	12.87
200	149	314	227	250	6200	236.6	13.73	2338.0	2398.35	922	14.94
250	186	393	283	300	9000	221.0709	13.01	2981.2	3001.76	944.98	16.37
325	242	510	368	400	16000	231.8	13.51	3823.6	3895.42	1041.39	19.23
425	317	667	481	500	25000	214.6	12.71	5082.0	5094.60	1261.1	21.82
525	391	825	595	600	36000	205.0	12.27	6301.4	6301.42	1453.64	22.63
625	466	982	708	800	64000	250.4	14.37	7139.1	7500.60	1503.64	28.68
Minimum		39 at HP=25	28 at HP=25	35 at HP =25	120 at HP=25	205.0 at HP=525	12.27 at HP=525	249.9 at HP=25	402.27 at HP=25	594.5 at HP=50	6.66 at HP=25 at arc-angle =342.0835
Maximum		982 at HP=625	708 at HP=625	800 at HP =626	64000 at HP=625	342.08 at HP=25	18.61 at HP=25	7139.1 at HP=625	7500.60 at HP=625	1504 at HP=625	28.6 at HP=625 at arc-angle =250.49

Table. 4

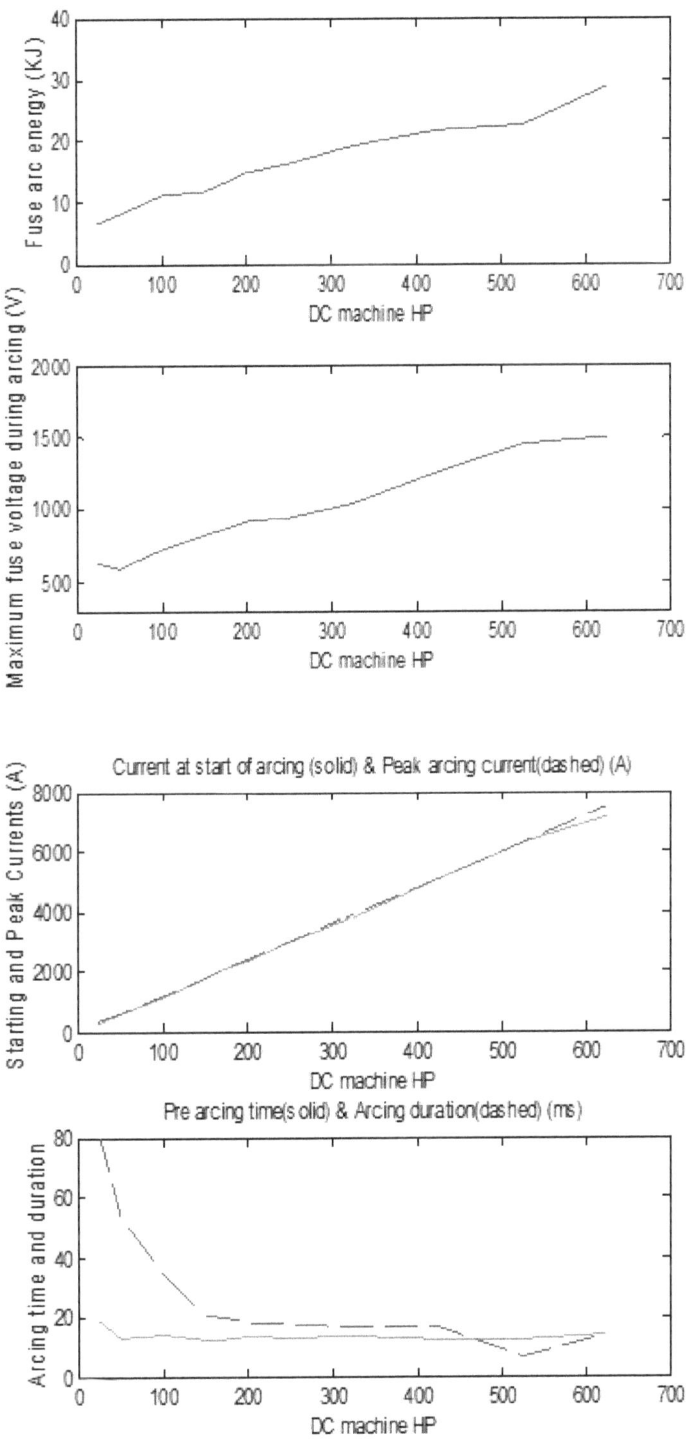

Fig. 8. Variation of the arc energy (KJ) and fuse voltage (V) , current at start of arcing (A), peak arcing current (A), pre arcing time(ms) and arcing duration (ms) versus the dc machine HP when 200 A fuse used as F1

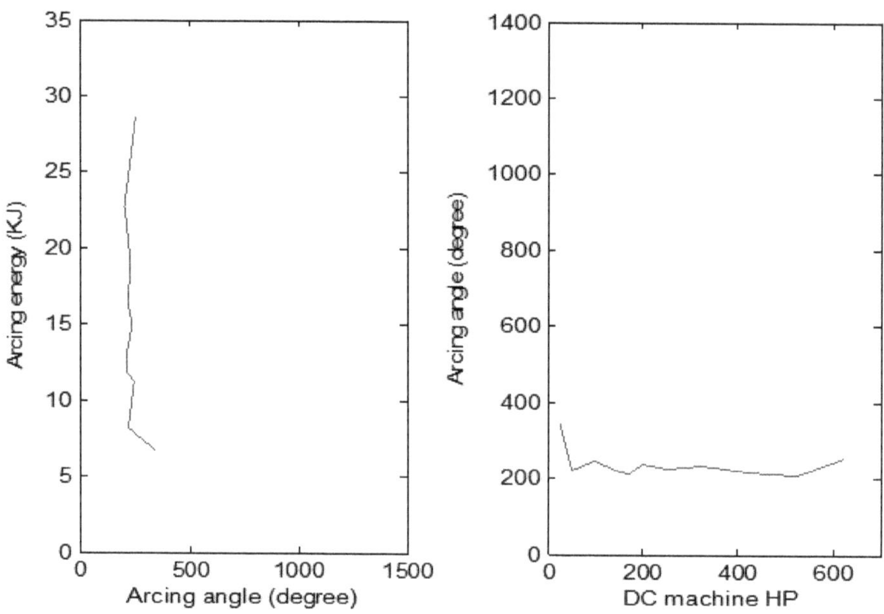

Fig. 9. Variation of the arc energy versus arc-angle and arc-angle versus the dc machine HP when the 200 A fuse used as F1

8. Conclusion

The paper investigated the effect of varying different parameters to the total response for fuses protecting against the regenerative circuit internal commutation fault. The effect of varying the motor time constant, supply impedance, number of fuses used to clear the fault and dc machine rating was studied. The model of 200A fuse is employed in this study and fuses in series with the semiconductor devices (F1) and fuses in ac line (F2) are both considered.

It was concluded that the location of the same fuse whether it is in series with the semiconductor devices (F1) or in the ac line will affect the loading condition of the dc machine I_{dc}. The percentage value of supply impedance and dc machine time constant play an important role for the total response of internal commutation fault. The stress on fuse within the range of mid dc machine time from 20msec to 50msec is not severe. However within the range 10-10msec and beyond 50 msec melting can occur in the critical melting zones which is severe condition. Varying the supply impedance will affect mainly the current at start of arcing and peak prospective current. If one fuse alone clears the circuit, the arc energy produced for internal commutation fault is high. When there are two fuses either sharing the breaking duty equally or the voltage is split in a ratio of 0.6:0.4, the arc energy reduced significantly. An adequate voltage rating for fuses protecting the regenerative dc drives is the ac line-line supply voltage. Increasing the dc machine and fuse ratings will increase the level of fuse voltage, starting and peak prospective currents and arc energy to higher level.

www.ingramcontent.com/pod-product-compliance
Ingram Content Group UK Ltd.
Pitfield, Milton Keynes, MK11 3LW, UK
UKHW061702190726
13853UKWH00008B/2353

9 798211 935136